DAY AND NIGHT

Library of Congress Cataloging-in-Publication Data

Butler, Daphne, 1945-
[Day and night]
First look at day and night / Daphne Butler.
p. cm. -- (First look)
Previously published as: Day and night. c1990.
Includes bibliographical references and index.
Summary: A simple introduction to day and night and the activities of each.
ISBN 0-8368-0505-4
1. Day--Juvenile literature. 2. Night--Juvenile literature. [1. Day. 2. Night.]
I. Title. II. Series: Butler, Daphne, 1945- First look.
QB633.B87 1991
525--dc20 90-10241

North American edition first published in 1991 by

Gareth Stevens Children's Books
1555 North RiverCenter Drive, Suite 201
Milwaukee, Wisconsin 53212, USA

Photograph credits: Robert Harding, 8, 20, 26; Camilla Jessel, 15; Mothercare, 29; ZEFA, cover, 7, 10, 11, 12, 16, 19, 21, 22, 23, 24

Series editor: Rita Reitci
Design: M&M Design Partnership
Cover design: Laurie Shock

Printed in the United States of America

1 2 3 4 5 6 7 8 9 97 96 95 94 93 92 91

DAY AND NIGHT

DAPHNE BUTLER

Gareth Stevens Children's Books
MILWAUKEE

Books in the
FIRST LOOK series:

FIRST LOOK AT
THE AIRPORT

FIRST LOOK AT
BOATS

FIRST LOOK AT
CARS

FIRST LOOK AT
CHANGING SEASONS

FIRST LOOK AT
DAY AND NIGHT

FIRST LOOK IN
THE FOREST

FIRST LOOK IN
THE HOSPITAL

FIRST LOOK
UNDER THE GROUND

CONTENTS

THE SUN SETS IN THE WEST

The sky is dark after sunset. But you can easily find the Moon and the first bright star of the evening.

Soon all the other stars will come out, too.

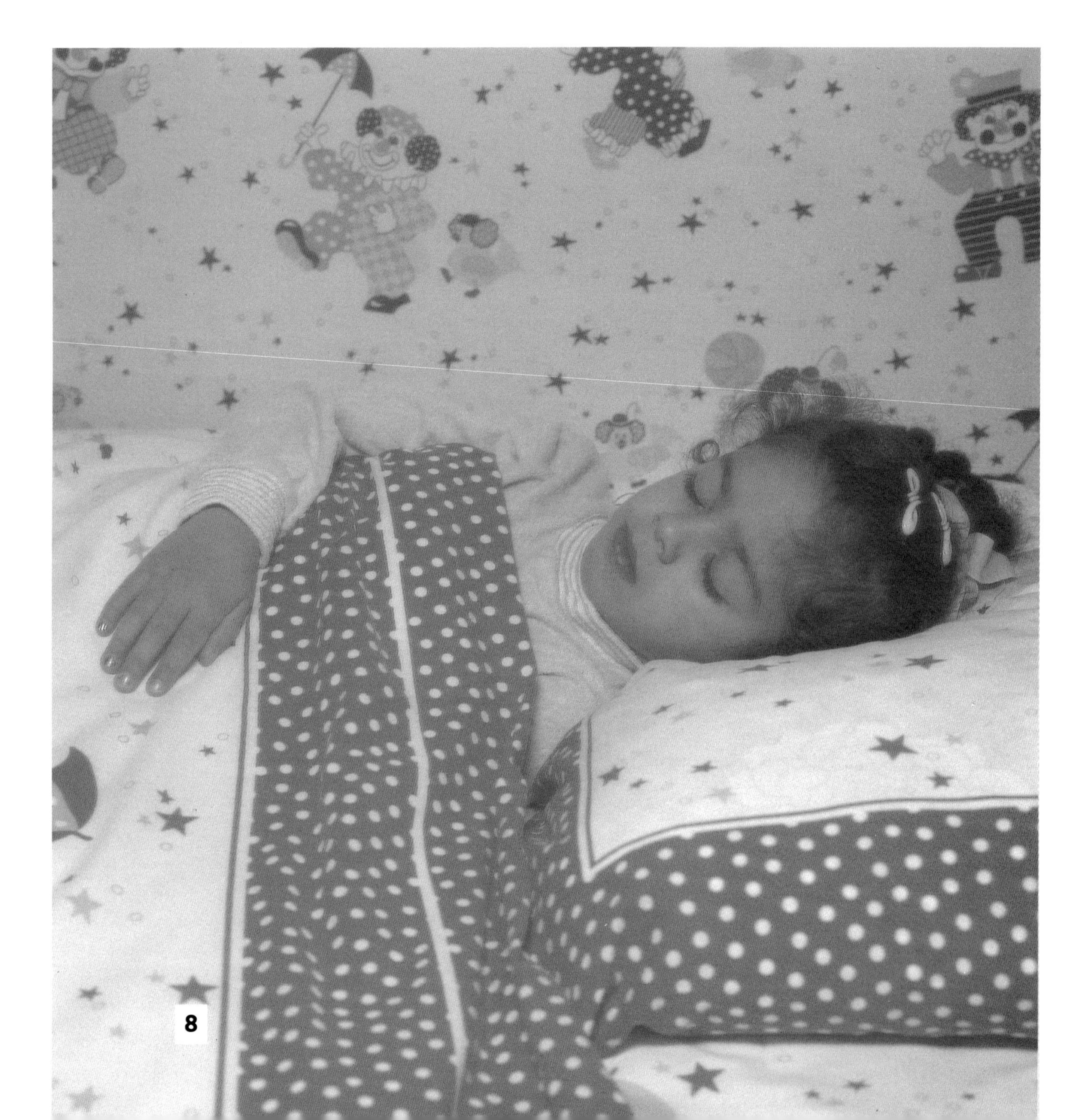

TIME FOR BED

Most people and many animals sleep during the night. These animals can't see very well in the dark. After a busy day, both people and animals need to rest.

The night is calm and quiet, and almost everybody is asleep.

ANIMALS GO HUNTING

Animals that hunt during the night can see well in the dark. They will sleep during the day.

Here are two animals that do this. Can you think of some others?

PEOPLE WORK AT NIGHT

People print newspapers at night. They run hospitals. And they work in gas stations.

Can you think of other places where people work at night?

TRAVELING AT NIGHT

Some planes make long journeys around the world. They fly all day and all night.

The passengers can sleep, but the crew must stay awake.

Have you ever traveled at night?

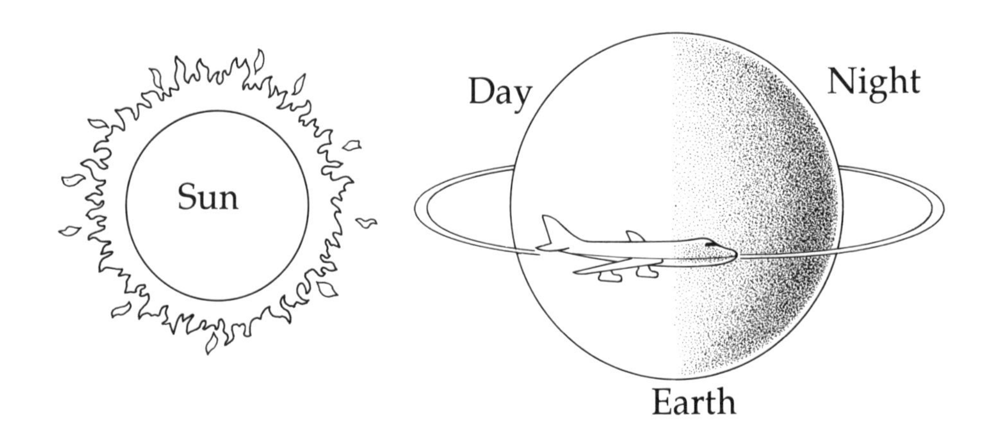

MONARCH
AIRLINES

16

DAWN BREAKS IN THE EAST

The Sun comes up over the horizon. Suddenly, it is daytime.

Birds begin to sing in the dawn light. The world starts to wake up.

MARKETS START EARLY

It takes a long time to set out the fruit and vegetables in a market. The stands must be ready when the first shoppers arrive.

When do you think the market people wake up?

19

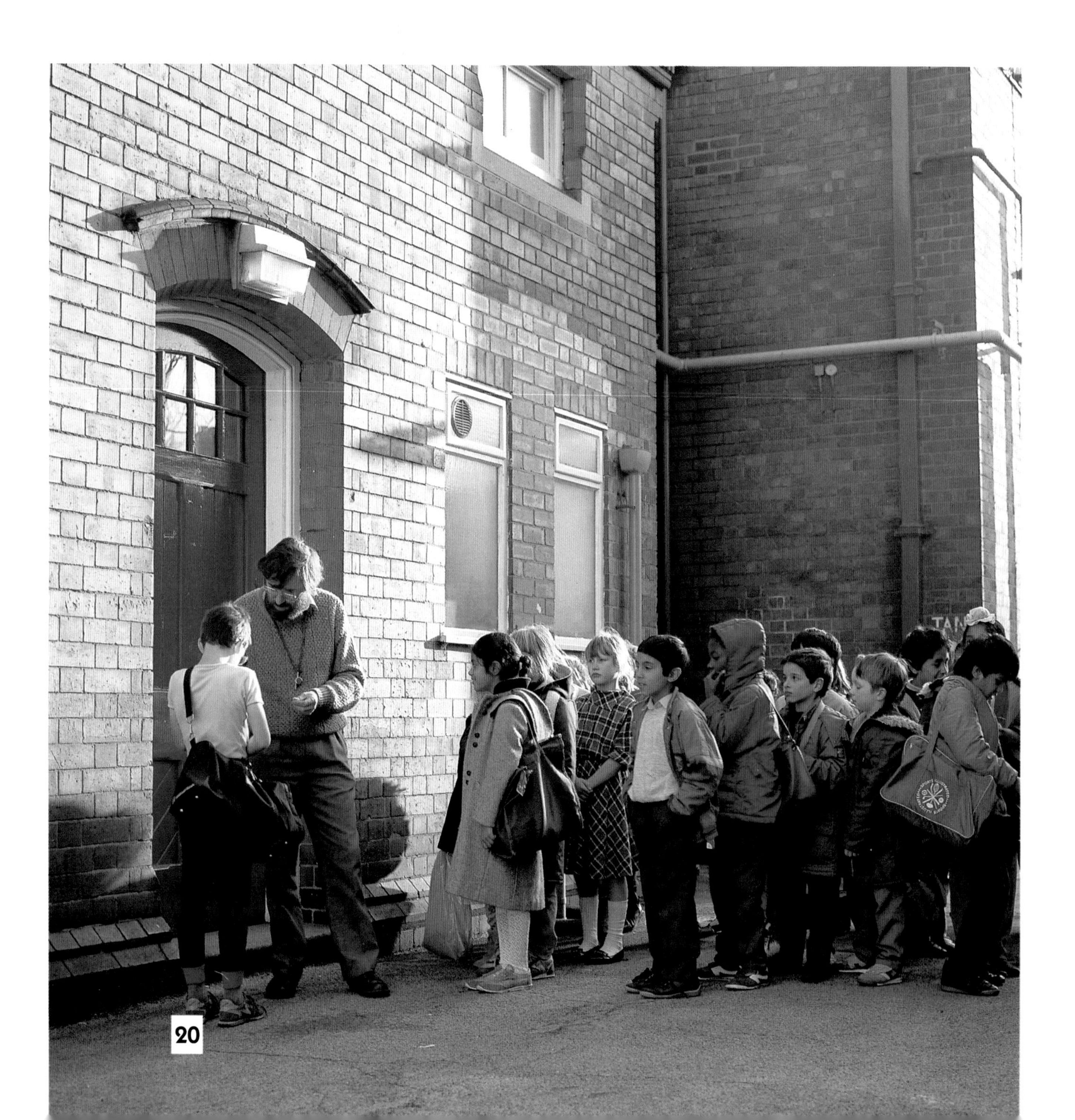

OFF TO WORK

Now it is morning. After breakfast, children go to school, and many adults go to work.

Some people walk, while others go by bicycle, car, bus, or train. What do you do?

ALL KINDS OF WORK

People do all kinds of different jobs. These are only a few of them.

Can you think of other jobs that people do?

TIME TO EAT

Most people have lunch in the middle of the day. They need a break before working in the afternoon.

At midday, the Sun is overhead.

SHADOWS GET LONGER

When the Sun shines, you can see your shadow. At midday, shadows are very small, but they grow longer in the afternoon.

How long are shadows in the morning? Can you use a shadow to tell the time?

NOW IT'S EVENING

Evening is the time to have fun, play a game, read a book, or watch television.

After the Sun sets, it's twilight. Soon it will be time for bed again. Why does night follow day and day follow night?

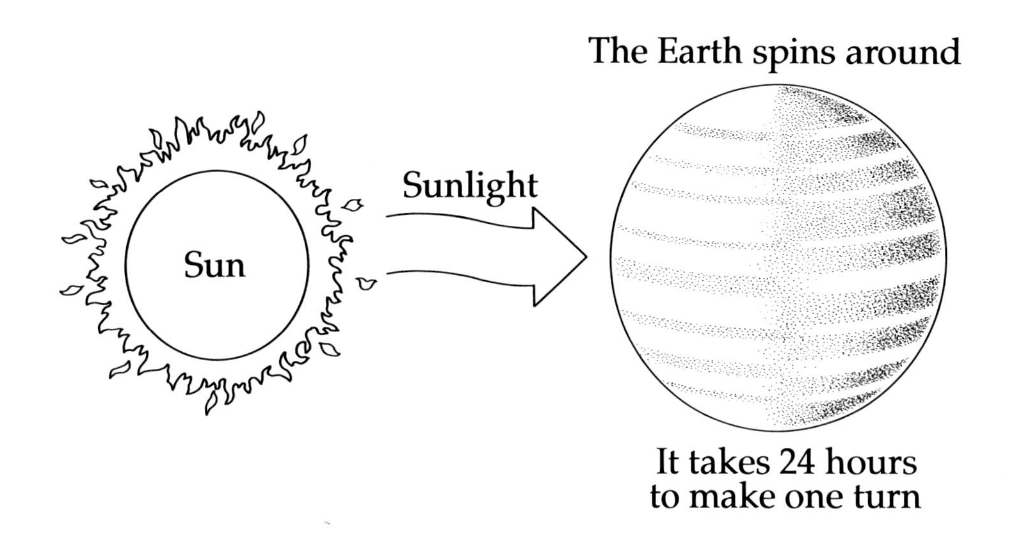

More Books about Day and Night

Animals at Night. Peters (Troll)
Clocks and How They Go. Gibbons (Harper & Row Junior Books)
Good Night, Night. Webber (Astor-Honor)
Let's Discover What People Do. (Raintree)
Midnight Animals. Tunney (Random House)
The Nighttime Book. Kunnas (Crown)
People at Work. Perham (Dillon)
People Working. Florian (Harper & Row Junior Books)
Richard Scarry's What Do People Do All Day? Scarry (Random House)
Shadow Magic. Seymour (Lothrop, Lee & Shepard)
A Summer Day. Florian (Greenwillow)
The Sun. Simon (Morrow)
Sun up, Sun Down. Gibbons (Voyager Books/Harcourt Brace Jovanovich)
Things People Do. Civardi (Usborne)
Time! Edmonds and Sachner (Gareth Stevens)
Walk When the Moon Is Full. Hamerstrom (Crossing Press)
The Way to Start a Day. Baylor (Macmillan)
What Time Is It Around the World? Baumann (Scroll Press)

Glossary

Afternoon: The later part of a day, from noon to sunset. Evening is the time between sunset and bedtime. Then it is night until the Sun rises.

Breakfast: The first meal of the day, usually eaten in the morning. People who work at night and sleep during the daytime may have their breakfast in the afternoon. This word is made up from the words *break* and *fast*. To break a fast means to end a time of not eating.

Crew: A group of people working together to get something done. Airplane crews work on planes that are flying. Ships have large crews to sail them. There are many other groups with special jobs — for example, repair crews, logging crews, and cleaning crews.

Dawn: The period of time when night begins to turn into day. We also call this time daybreak.

Day: The time it takes the Earth to make one full turn. This is 24 hours. We also say a day is the 24 hours from midnight to midnight. *Day* can also mean just the period of light between sunrise and sunset.

Horizon: The line where Earth and sky seem to meet. This line will be different for every place from which you look at it.

Morning: The first part of a day, from dawn to noon. This word can also mean the time from midnight until noon.

Twilight: The period of time from sunset to dark. There is still some sunlight, even though the Sun has already gone past the horizon. The dim part of twilight, when it is turning into night, is called dusk.

Index

A number that is in **boldface** type means that the page has a picture of the subject on it.